Reading Essentials®
in Science

LIVING WONDERS

Genetics

SUSAN GLASS

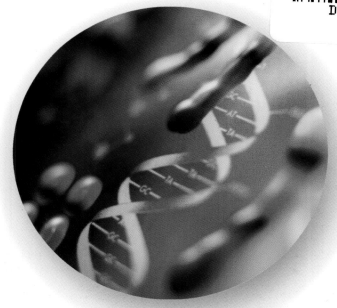

PERFECTION LEARNING®

Editorial Director:	Susan C. Thies
Editor:	Mary L. Bush
Design Director:	Randy Messer
Book Design:	Michelle Glass
Cover Design:	Michael A. Aspengren

A special thanks to the following for their scientific review of the book:

Paul Pistek, Instructor of Biological Sciences, North Iowa Area Community College

Jeffrey Bush, Field Engineer, Vessco, Inc.

Image Credits:

Copenhagen Library: p. 7; Associated Press: pp. 36, 44; ©Bettmann/CORBIS: pp. 11, 13; ©George Shelley/CORBIS: p. 15; ©Ted Spiegel/CORBIS: p. 17; ©Robert Pickett/CORBIS: p. 20 (top); ©Nkajlah Feanny/CORBIS SABA: p. 41; ©Roger Tidman/CORBIS: p. 35

Corel Professional Photos: pp. 4–5 (middle); Photos.com: front cover, back cover, pp. 1, 2, 3, 4, 5 (right), 9, 10, 14, 16, 20 (bottom three images), 21, 22, 23, 24, 25 (right), 26, 27, 28, 29, 30, 31, 33, 34 (bottom), 37, 39, 42, background on all sidebars; Ingram Publishing: pp. 5 (top), 32 (bottom), 43; ©Royalty-Free/CORBIS: p. 32 (top), backgrounds on all spreads; ArtToday: pp. 6, 8; Michael A. Aspengren: pp. 18, 19, 34 (top), 40; Photo compliments of Melissa Hoelker: p. 25 (top)

For information, contact
Perfection Learning® Corporation
1000 North Second Avenue, P.O. Box 500
Logan, Iowa 51546-0500.
Phone: 1-800-831-4190
Fax: 1-800-543-2745
Reinforced Library Binding ISBN-13: 978-0-7569-4480-3
Reinforced Library Binding ISBN-10: 0-7569-4480-5
Paperback ISBN-13: 978-0-7891-6232-8
Paperback ISBN-10: 0-7891-6232-6
3 4 5 6 7 8 PP 12 11 10 09 08 07
perfectionlearning.com

CONTENTS

chapter one

Who Do You Look Like?

An Introduction to Heredity and Genetics

Has anyone ever told you that you have your mother's eyes or your father's ears? Or maybe you have your weird Uncle Wilbur's nose. Why do you look like these people? The answer lies in **heredity**.

Every living thing has **traits**. Traits are characteristics that get passed on to **offspring**. Eye color is a trait in animals. Leaf shape is a trait in plants. A long neck is a trait that giraffes give to their offspring. Chameleons pass on their ability to change color. Behaviors like nest building or flying are also traits. The system of passing these traits on from one generation to the next is called *heredity*.

Genetics is the branch of science that studies heredity. Scientists who study genetics are called *geneticists*. Geneticists work to understand how living things pass on their traits and how this affects the variations, or differences, in all life-forms.

A BRIEF HISTORY OF HEREDITY AND GENETICS
● ● ●

Deer don't give birth to donkeys, and donkeys don't give birth to ducks. People have always realized that offspring are similar to their parents. Different cultures have had different ideas about how this happens.

In ancient Greece, philosophers (thinkers) used reasoning to try to figure out how heredity worked.

Aristotle was the greatest of these philosophers. He and other Greek thinkers didn't understand how babies inherited traits from their parents. They noticed that children weren't exact copies of just one parent. They seemed to get a combination of traits from both of their parents. Aristotle reasoned that children were made when something he called "the substance" from the mother was magically influenced by something called "the form" from the father. Aristotle eventually came to believe that blood was the basic unit that passed characteristics from parent to child. Although he died in 322 B.C., his ideas lived on for two thousand years.

Blood Relatives

Aristotle's ideas about blood and heredity influenced expressions still used in our language today. The phrases "blood relatives," "bloodlines," and "related by blood" all stem from Aristotle's theories.

The understanding of heredity did not move forward very far until an amateur scientist named Anton van Leeuwenhoek invented the microscope in the 1600s. Finally scientists could see the **cells** that pass on characteristics from parent to child. Later, using bigger and better microscopes, scientists explored a part of the cell known as the **nucleus**. Most cells have a nucleus that contains **chromosomes**. Scientists discovered that it's these chromosomes that hold the units that pass traits from parent to child.

In 1909, a scientist named Wilhelm Johannsen identified these units and named them **genes**. Genes are the basic units of heredity. Aristotle was wrong. It is genes—not blood—that pass traits down through generations.

MODERN GENETICS

● ● ●

Modern genetics is an exciting field of science. Stunning developments have shown uses for genetic research that could never have been imagined by Aristotle or his followers. Scientists are busy learning how to cure inherited diseases, how to develop plants that are insect-proof, and even how to clone animals. The study of genetics has revolutionized crime fighting. Analyzing samples of blood, hair, saliva, or skin can prove guilt or innocence beyond a shadow of a doubt.

Wilhelm Johannsen

Studying genetics has led to so many advances in recent years that it's hard to believe that it all began with a little-known monk who grew pea plants in his spare time.

chapter two

The Father of Genetics

Gregor Mendel

Gregor Mendel is now considered a scientific giant and the father of genetics. But when he was alive, nobody paid much attention to his work. In fact, they ignored it. The importance of his work was not recognized until long after Mendel died. Thirty-five years after he published his findings in scientific journals and sixteen years after he died, scientists working on the laws of heredity stumbled across his research. At last Mendel got the credit he deserved, and his work became the foundation for understanding heredity.

MENDEL'S MYSTERY

• • •

Johann Mendel was born in 1822 into a poor Austrian family. He took the name Gregor when he became a monk. Mendel studied math and science. He taught high school, even though he was officially considered a substitute teacher because he could not pass the exam to be a regular science teacher. When he wasn't teaching, he looked after the monastery's garden.

Some of his fellow monks were crop **breeders**. Soon Mendel started experimenting to help them improve their crops. He became an amateur botanist and set out to solve a scientific mystery. How do parent plants pass traits to their offspring?

About Botanists

A botanist is a scientist who studies plants.

Mendel began his investigations with pea plants in 1856 and kept experimenting for several years. When he was promoted to head of the monastery in 1868, he no longer had time for his plants. But over those 12 years of systematic experimenting, Mendel kept careful records of thousands of pea plants to check for patterns in the traits being passed from parent plants to offspring plants. Many years later, those records were recognized as the beginnings of the science of genetics.

MENDEL'S EXPERIMENTS

● ● ●

Mendel was lucky that he chose pea plants for his experiments. Unlike a lot of other plants, peas have many traits that only come in two forms. Their stems are either tall or short, never in between. Their seeds are either wrinkled or smooth and either yellow or green. The seeds come in pods that are either smooth or bumpy and either green or yellow. The pea flowers are either purple or white and grow either at the end or the sides of stems. Offspring inherit each of these traits in either one form or the other. There is never a mix of traits or any new forms.

Mendel studied these seven different traits in pea plants. At first, he focused on one trait at a time. Since each of the traits has only two different forms, he could easily see patterns in how the traits were passed from one generation to the next.

Mendel started his experiments with purebred plants. A purebred plant is one that always produces offspring with the same form of a trait as the parent plant. For example, a purebred tall pea plant will always produce a tall offspring. A purebred yellow-seeded pea plant will always produce new plants that have yellow seeds.

For his first experiment, Mendel crossed purebred tall plants with purebred short ones. He called this first group of plants the P generation because they were the parent generation. He called their offspring the F1, or first filial generation. All of the new plants in the second generation were tall. The shortness trait seemed to have disappeared.

Mendel crossed the F1 generation to produce offspring. Mendel called the next generation F2 (second filial generation). This generation was a mix of short plants and tall plants. The shortness trait had not disappeared after all! When Mendel counted them, he discovered that three out of four plants in this generation were tall. One out of four was short.

Mendel decided to call the tall form the **dominant** form, since it seemed to dominate, or overpower, the short form. He called the short form the **recessive** form. What does *recessive* mean? The root word is *recess*. You know recess. You love recess. The word *recess* means "a temporary going away or a temporary removing." When you go out for recess, you temporarily remove yourself from the classroom. A recessive form of a trait seems to go away, only to come back in later generations.

An illustration of Gregor Mendel at work in his garden

All in the Family

Filial means "son" in Latin. So filial generations were the first generations of "sons."

Mendel also
experimented with the six other traits.
He crossed plants with opposite traits the same way he
had done for plant height. Surprise! The results were the
same as those in the height experiment. Only one form of the
trait appeared in the first generation. In the second generation, the
missing form always reappeared in about one-fourth of the plants. For
example, when he crossed plants with green seeds and plants with
yellow seeds, the first (F1) generation all had yellow seeds. The green-
seed trait seemed to have disappeared. But it was only hidden—or out
to recess, so to speak. It came back in the second generation of plants.
One out of four of the F2 pea plants had green seeds.

MENDEL'S CONCLUSIONS
● ● ●

Mendel studied his results and decided that individual "factors"
must control the passing on of these traits. These factors seemed to be
in pairs. He concluded that the female parent plant must contribute one
factor and the male parent plant must contribute the other factor. What
Mendel called *factors* are now called *genes*.

In his records, Mendel labeled each factor with a letter. He used a
capital letter to stand for a dominant trait. He used a small letter to
stand for a recessive trait. A capital T stood for tallness in pea plants. A
lowercase t stood for the recessive trait of shortness. If a plant had two
tall genes (TT), it was tall. If it had
two shortness genes
(tt), it was short.
If a plant had
one tall and one
shortness gene (Tt),
the dominant trait overpowered the
recessive one and the plant was tall.

Purebred or Hybrid?
Organisms with genes that are the same for a
particular trait, such as TT or tt, are called *purebreds*.
Living things that have genes that are different for a
trait, such as Tt, are called *hybrids*.

Mendel studied about 30,000 pea plants. He found that their heredity followed certain patterns. He recognized that traits are inherited in units, one from each parent. He learned that some traits are dominant and others are recessive. He also realized that hybrids produce some offspring with dominant traits and some offspring with recessive traits. Mendel's conclusions later became known as Mendel's Laws of Heredity.

Punnett Squares

Punnett squares are a type of chart used to show the possible combinations of genes that two parents could pass on to their children. They also show the probability for each gene combination and expressed trait. The squares were named for Reginald Punnett, an English geneticist who worked with chickens to figure out how to link traits to males or females. This is a Punnett square showing the possible gene combinations for two tall plants that carry a recessive shortness gene (Tt).

	T	t
T	TT (tall)	Tt (tall)
t	Tt (tall)	tt (short)

AFTER MENDEL
● ● ●

Mendel's work was forgotten for 34 years. Then in 1900, three different scientists made many of the same discoveries that Mendel had made. These men were Hugo de Vries, Carl Correns, and Erich Tschermak. When the three men studied scientific journals to see what others had learned before, they discovered Mendel's work. They were amazed to learn that Mendel had beaten them to the punch decades earlier.

Hugo de Vries

Carl Correns went further than Mendel and discovered that in some gene pairs, neither gene is dominant or recessive. When neither trait is masked or hidden, the traits are **codominant**. In cattle, for example, white hair and red hair are codominant traits. If a cow gets a gene for red hair and a gene for white hair, it ends up with a mix of both. From a distance, these cows are a pinkish brown, or roan, color. In Erminette chickens, black feathers and white feathers are codominant traits, so these chickens end up covered partly in black feathers and partly in white ones.

Since Mendel's time, many scientists have made contributions to the study of genetics. All of these scientists' conclusions, however, are based on the laws of heredity that resulted from Gregor Mendel's pea plants.

Try This!

To get an idea of how dominant and recessive traits work, gather ten brown M&Ms, ten blue M&Ms, and two small paper bags. The brown M&Ms are dominant genes for M&M color. The blue M&Ms are recessive genes for M&M color. Place five blue and five brown M&Ms in each bag. Draw an M&M out of each bag and record their colors. Use a capital B for brown and lowercase b for blue. Based on Mendel's laws, decide what color a new offspring M&M would be given the combination of genes. For example, if you draw one brown M&M and one blue M&M (Bb), then the new M&M would be brown. Draw out combinations until all the M&Ms are gone. How many brown M&Ms do you have? How many blue ones?

chapter three

The Keys to Genetics

DNA, Genes, and Chromosomes

To understand the science of genetics, you must understand three key concepts— **DNA**, genes, and chromosomes.

DNA
● ● ●

DNA is short for deoxyribonucleic acid. This amazing **molecule** is found in every living thing. It is responsible for the formation, growth, and reproduction of individual cells and entire organisms.

Each person's DNA is unique. Only identical siblings, such as twins and triplets, have the same DNA.

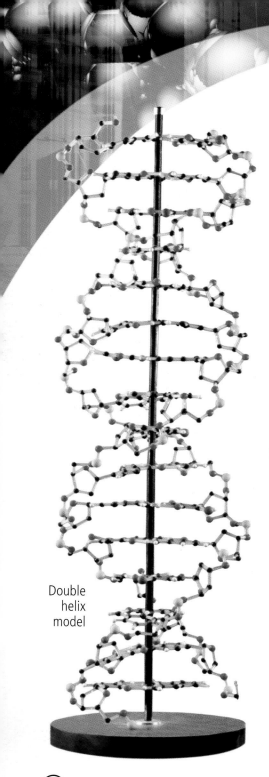

Double helix model

For many years, scientists puzzled over how something as tiny as a DNA molecule could hold all the information needed for the development of something as complicated as a human, a whale, or even a maple tree. The race was on to solve the puzzle.

In 1953, two scientists, Francis Crick and James Watson, finally figured out the basic twisted-ladder shape of DNA. They called the spiraling shape a "double helix." A helix is a spiral that twists like a piece of string wrapped around a pencil.

DNA consists of two spirals of molecules twisted together. The spirals are connected by four **bases**—adenine, thymine, guanine, and cytosine. These are commonly referred to as A, T, G, and C. When paired up, these bases form the rungs of the DNA ladder shape. The amount of each base and the order in which they are arranged is different in every organism.

Perfect Pairs

When bases pair up to form the DNA ladder, A (adenine) always attaches to T (thymine) and C (cytosine) always attaches to G (guanine).

A Prize-Winning Discovery

Francis Crick (left) and James Watson (right) were awarded the Nobel Prize for Medicine in 1962 for their work with DNA.

Once the structure of DNA was figured out, scientists began to look at how it works. DNA molecules can copy themselves. When cells divide, the new cells get copies of the same genetic instructions, or genetic code, as the original cells.

DNA Goes the Distance

The human body has trillions of feet of DNA. If the DNA in just one human cell was completely uncoiled and stretched into a straight line, it would be about the height of an adult (five to six feet). If you could uncoil and straighten out all of the DNA in your body, it would stretch to the Sun and back many times.

DNA helps in the formation of **proteins** by transferring genetic information. Proteins are found in the cells of all living things. They direct the day-to-day work of the cells. Proteins are needed for growth, repair, and replacement of cells. They also help build muscles and organs. It is estimated that humans have about one million different proteins. Each one has a special job.

DNA is found in each cell's nucleus. But proteins aren't manufactured in the nucleus. They get made in the tiny **ribosomes** found outside the nucleus. In order for proteins to be made, a messenger must take the genetic code from the DNA in the nucleus to the **cytoplasm** of the cell. That carrier is called messenger RNA. Messenger RNA copies information from DNA in the nucleus and carries it to the cytoplasm. There the RNA attaches to a ribosome. The ribosome reads the code and produces a protein according to the instructions.

Just in Case You Were Wondering . . .

RNA stands for ribonucleic acid.

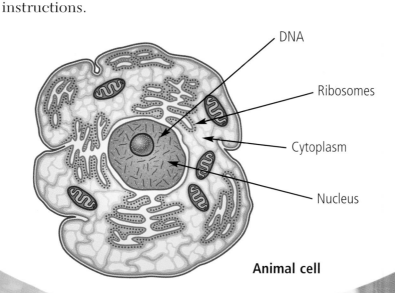

DNA

Ribosomes

Cytoplasm

Nucleus

Animal cell

Try This!

Make a short segment of DNA using six toothpicks, paint or markers, and two pieces of licorice. Choose four colors, one to represent each of the four bases—A, T, G, and C. Paint or color halves of toothpicks with corresponding colors. Remember that A and T or G and C go together. For example, if A is yellow and T is blue, then several of your toothpicks will be half yellow and half blue for the A and T base pairs. Since each DNA molecule is unique, you can decide how many toothpicks will be A and T pairs and how many will be G and C pairs. When all six toothpicks are done, stick the ends of the toothpicks into the sides of the licorice pieces, forming a ladder shape.

GENES

• • •

Genes are made up of DNA. Each species of living things has a certain number of genes in its cells. It is estimated that humans have about 20,000 to 30,000 different genes, while fruit flies have only about 13,000.

Genes have two main jobs. Genes are like blueprints, or instructions, that tell cells how to make particular proteins. Scientists believe that each gene in the body may be able to make many different kinds of proteins. These proteins are what determines an organism's traits. Whether you have blue or brown eyes, black or blond hair, and big or small feet depends on your genes. Genes control whether a plant has round or pointed leaves, a tall or short stem, and red or yellow flowers.

The second job of genes is to pass on traits so that each species keeps its unique characteristics. Humans have different traits than ducks. Both have different traits than maple trees. These different traits are what makes each type of organism special.

CHROMOSOMES

● ● ●

Genes are found on chromosomes. Chromosomes are threadlike structures found in the nucleus of a cell in most organisms. Each chromosome may have thousands of genes lined up on it. In organisms that have pairs of chromosomes, one set came from the female parent and one set came from the male parent.

Each type of organism has a different number of chromosomes. The number of chromosomes has nothing to do with the size or importance of the organism. For example, humans have 46 chromosomes (23 pairs), while goldfish have 96 (48 pairs).

Check Out the Chromosomes!

Organism	Number of Chromosomes
dog	78
chicken	78
gorilla	48
chimpanzee	48
mouse	40
fruit fly	8
fern	512
tomato	24
corn plant	20

chapter four

Reproduction

How Do Traits Get Passed On?

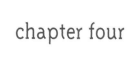

It's sad but true that living things don't live forever. So plants, animals, and all other organisms must reproduce for their species to survive. **Reproduction** is the process in which living things make more living things like themselves. The ability to reproduce is one characteristic that separates living from nonliving things. Plants make more plants. Animals make more animals. Rocks don't make baby rocks.

Simple one-celled organisms reproduce by dividing themselves into two new cells. Bacteria divide to make more identical bacteria. Protozoa can divide to make new identical protozoa.

Many organisms reproduce in one of two ways—asexual or sexual reproduction.

Protozoa

ASEXUAL REPRODUCTION
● ● ●

In asexual reproduction, a new organism is created from a single parent. There are several forms of asexual reproduction. Some organisms simply split in two to create offspring. Others have parts that break off and develop into new organisms. Flatworms, crabgrass, and yeasts reproduce when part of the parent breaks away and becomes a new organism. Some parent plants produce a special structure used to make an offspring. The strawberry plant creates a skinny growth called a *runner* that spreads across the ground. Where the runner puts down roots, a new strawberry plant grows.

Since asexual reproduction requires only one parent, the offspring receives an exact copy of that parent's genes. The offspring will be an exact copy, or clone, of the parent.

SEXUAL REPRODUCTION

• • •

Most plants and animals reproduce by sexual reproduction. In sexual reproduction, two parents contribute to the offspring. The male parent contributes a male reproductive cell called a *sperm*. The female parent contributes a female reproductive cell called an *egg*. Each of these special cells has only half of the chromosomes normally found in the organisms' cells. When the two reproductive cells unite, the new organism now has a complete set of chromosomes. Because the chromosomes are a mix from two different parents, the offspring shares traits from both parents but is not identical to either one.

In some cases, an individual organism may be both male and female. These organisms are called *hermaphrodites*. Hermaphrodites produce both eggs and sperm. Sometimes hermaphrodites must reproduce with other hermaphrodites. Other times they can mate with themselves. Hermaphrodites are common in plants. Mendel's pea plants were hermaphroditic.

In most plants, the joining of a sperm and an egg produces a seed. The seed contains half of the female plant's genes and half of the male plant's genes. When the seed grows into a new plant, the plant has a combination of traits from the two parent plants.

Try This!

Soak some dried pinto or lima beans in water overnight. After soaking, take off the outer protective layer called the *seed coat*. Split the seed open. Can you find the **embryo**? The embryo is a baby plant. The rest of the seed is food for the plant. When an embryo starts to grow, it turns into a seedling, or child plant. Eventually it will grow into an adult plant that can make its own food.

Unlike plants, the joining of a sperm and egg in animals does not result in seeds. The "baby" offspring usually resembles an adult but in a much smaller, immature form. Most animal babies remain inside the mother or inside an egg until they are ready for life on their own.

Regardless of the form of the developing offspring, the key to sexual reproduction is the joining of two special reproductive cells. This ensures a new life with a mix of genes from both parents. This mix gives the offspring a set of unique traits.

Flamingos usually lay just one large egg at a time.

A Special Exception

Someone always has to break the rules! Several species of whiptail lizards, including the New Mexico whiptail, are all females. In these species, the mother whiptail lizards produce eggs without contributing male reproductive cells. All of the young are exact copies of their mothers.

MALE OR FEMALE?

• • •

A special pair of chromosomes in humans and most other animals determines whether an offspring will be a male or female. These chromosomes are called the sex chromosomes. Females have two X sex chromosomes. Males have one X sex chromosome and one Y sex chromosome. When a female passes on half her chromosomes to an offspring, she passes on one of the X chromosomes. When a man passes on half of his chromosomes, he either contributes the X or the Y chromosome. When the chromosomes pair up in the new offspring, it receives either two X chromosomes or an X and a Y chromosome. Two X chromosomes means the offspring will be a female, while an X and a Y will result in a male. So you can thank your dad for determining whether you are a boy or a girl!

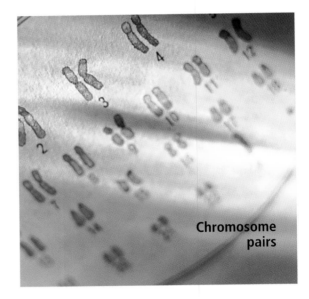

Chromosome pairs

chapter five

A Heads-Up on Heredity

Understanding Human Traits

Human heredity isn't nearly as simple as Mendel's pea plants. Humans have thousands of genes. An inherited trait, such as eye color, ear shape, skin color, etc., can be determined by one gene or by many. Genes determine if you are right-handed or left-handed. They make you red-haired, blond-haired, or no-haired (bald). Genes can even decide if you can curl your tongue like a taco shell.

Most human traits are caused by more than one gene. The genes work together as a group to produce a single trait. Your height is produced by many genes. Your skin color came from four pairs of genes working together. Eye color is also the result of more than two genes.

ALLELES

● ● ●

Some human traits are determined by a single gene with two **alleles**. An allele is a different form of a gene. Remember the pea plants? Tallness and shortness were two alleles for the pea plant height gene. The dominant allele's trait always shows up in the organism. The recessive allele, on the other hand, is hidden whenever it is paired with a dominant allele.

Simple single-gene traits work the same way in humans. The traits that are carried on a single gene with two alleles either appear in a person or don't. Just like with the pea plant, there's no mix of the traits. Dimples are an example of this. You either have dimples or you don't. There are no "half" dimples or "sort of" dimples.

A widow's peak is another allele trait. A widow's peak is a hairline that dips down to a point above the middle of the forehead. You either have a widow's peak or you don't. If you have two recessive alleles for a widow's peak, you don't have one. If you inherit one or two dominant widow's peak genes from your parents, then you inherit the widow's peak.

The chart to the right shows some examples of dominant and recessive single gene traits.

Dominant Traits	Recessive Traits
tongue curling	noncurling tongue
full lips	thin lips
freckles	no freckles
farsightedness	normal vision
connected eyebrows	separate eyebrows
straight thumb	hitchhiker's (bent) thumb
unattached earlobes	attached earlobes
naturally resting left thumb on top when folding hands together	naturally resting right thumb on top when folding hands together

Some traits in humans are controlled by a single gene that has more than two alleles. Such genes have three or more forms of a gene that can cause one trait. Each person, though, can carry only two of those alleles. (Your chromosomes only come in pairs!) The combination of the two you have determines what traits you have (or don't have).

Blood types work this way. There are three alleles for the gene that determines blood types—A, B, and O. Two of them are codominant (A, B) and one is recessive (O). With the mixing and matching possibilities, we end up with four main blood types—A, B, AB, and O. A person with a dominant A and recessive O allele has a blood type of A. A person with codominant A and B alleles has AB blood. Only a person with two O alleles has an O blood type.

X AND Y TRAITS

Some traits come from genes found on the X and Y sex chromosomes. These are called sex-linked genes. The X does not carry the same set of genes as the Y. Because a male only has one X chromosome, whatever allele is carried on it will determine what trait appears. Females receive two X chromosomes, so the combination of alleles carried on both chromosomes will determine the female's traits. Because of this, males have recessive sex-linked traits more than females. Colorblindness and hemophilia (a bleeding disorder) are two sex-linked traits.

HEREDITY VERSUS ENVIRONMENT

After all this talk about heredity, genes, chromosomes, and traits, does it surprise you to learn that not all of your characteristics are inherited? While heredity does determine many of your mental and physical traits, the **environment** also plays a role in how you actually think, feel, and look.

While you inherit your parents' genes for "smartness" and various skills, how you develop your intelligence and skills will make a difference in the role the traits play in your life. A child who reads a lot, practices math skills, and participates in a variety of learning activities will generally be more successful than one who doesn't. Someone with inherited musical talent will go much further if he or she has the opportunity to take lessons, practice, perform, etc. You may have your mother's or father's athletic talents, but if you never pick up a ball, practice, or play games, it won't do you much good.

Your height and weight tendencies are inherited. However, your diet can affect your actual height and weight. A healthy diet, for instance, can make you taller than a junk food diet. A lot of fast food might make you heavier than if you eat a balanced, low-fat diet.

Physical traits can change with the choices you make. You inherited the muscles in your arms. If you lift a lot of weights, you can make them bigger. But if you never lift anything heavier than a TV remote, your muscles might be smaller than they could be. You might have inherited pale skin, but spending a lot of time in the Sun may result in darker skin.

Some diseases are inherited. There's nothing you can do about it if you're unlucky enough to get the gene that passes them on. However, some genes only give you a tendency, or higher likelihood, of getting certain diseases. High blood pressure and heart disease tend to run in families. Your environment and the choices you make may affect your chances of actually developing the diseases. Eating a balanced diet, maintaining a healthy weight, exercising regularly, and finding positive ways to deal with stress may prevent you from getting these diseases.

Pairs of Proof

Studies done on identical twins who were raised in different homes have proven the heredity versus environment theory. While identical in gene makeup, the twins developed differences in their actual characteristics based on their surroundings and lifestyles.

The Same Is True for Plants

Heredity and environment interact in other organisms too. For example, a plant might have a gene for tallness, but if it does not get enough water or minerals, it won't grow to its full size. Plants with genes to produce flowers may not actually develop the flowers if they don't receive sunlight or food.

chapter six

Mutations

Not Just in the Movies

Movies such as *The Teenage Mutant Ninja Turtles* and the *X-Men* are based on **mutant** characters with traits that make them different from the rest of the population. But did you ever stop to think about how these characters got their name or their strange characteristics? From the science of genetics, of course!

Primrose

WHAT IS A MUTATION?
● ● ●

In 1886, Dutch botanist Hugo de Vries made an accidental discovery. Like Mendel, de Vries grew plants, but he used evening primroses instead of peas. His results were much like Mendel's. Then something strange happened. Every once in a while, a new variety of primrose would suddenly appear. This flower wasn't like any that he had seen before. De Vries called these sudden changes **mutations**.

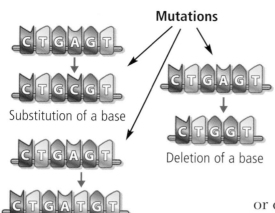

Mutations

Substitution of a base

Deletion of a base

Insertion of a base

When cells divide and the genes are copying themselves, the bases in the DNA ladder may match up incorrectly or in the wrong order. Sometimes the chromosomes don't separate perfectly, and a cell can end up with too few or too many chromosomes. Any change in a gene or chromosome is called a *mutation*.

Mutation comes from an old Latin word that means "change."

HOW DO MUTATIONS AFFECT CELLS?
● ● ●

Mutations can and do happen occasionally in an organism's cells. This is usually no big deal if it happens in a muscle cell, a skin cell, or a leaf cell. Most of these mutations aren't noticeable, and many are repaired before they can continue reproducing. And because they are in nonreproductive body cells, the mutations are not passed on to future generations.

But if a mutation happens in a reproductive cell, the mutation will be passed on. For example, if a female pig has a mutation in an egg cell that is fertilized and grows into a piglet, the piglet will carry the mutation. Its genetic blueprint now has something new that was not there before in its family tree. And that new mutation can be passed on to the piglet's offspring and their descendants.

HARMFUL OR HELPFUL?

● ● ●

Peppered moth

Mutation may sound like a scary thing. You might picture mutants as creatures with three eyes or super strength. But mutants are not the monsters you see in movies. In reality, mutations are very common in the life of cells. Most have little or no effect on the organism. Some are even helpful to an organism's survival.

Hugo de Vries was one of the first to realize how important mutations are in the **evolution** of life. Mutations create variety in plants and animals. Variety in a species helps that species adapt and survive when conditions change.

An example of this occurred in England in the 1800s. A species of moths called *peppered moths* lived in the country. Almost all of the moths were light-colored. These moths hid themselves from predators on light-colored tree trunks. Only a few peppered moths had a mutation that made them black. When factories began burning so much coal that the soot covered the trees and made them black, the light-colored moths were quickly eaten by birds. The few black moths, however, survived because they were now hidden in the soot-covered trees. These mutant black moths reproduced, and in time, most of the peppered moths were black. Later, when pollution laws reduced the amount of soot in the air, the trees became lighter and the number of light-colored moths increased again. Without mutations, the peppered moth would have likely died off.

Some mutations *are* harmful. They cause diseases or defects that threaten a person's health. Cystic fibrosis is an illness caused by a gene mutation. People with cystic fibrosis have difficulty breathing and need treatments to clear their lungs.

Sometimes chromosomes mutate and an offspring receives too few or too many pairs of chromosomes. In humans, an extra pair of chromosome 21 causes Down Syndrome. Children with Down Syndrome have physical and mental developmental difficulties. They also share physical traits, such as shortness, slanted eyes, and poor muscle tone.

In addition to spontaneous-occurring mutations, geneticists have learned that mutations can be caused by sources outside of the body. X rays, ultraviolet light from the Sun, and chemicals such as those found in cigarette smoke can cause cell mutations. These environmental factors are called *mutagens*.

Opening ceremonies of the 32nd annual Special Olympics Summer Games

chapter seven

It's All in the Genes!

Recent Developments in Genetics

From Gregor Mendel's pea plants to **cloning** a sheep, genetics has continued to develop into an interesting branch of science based on the amazing world of genes.

CRIME-SOLVING DNA

● ● ●

DNA is solving more crimes these days than Sherlock Holmes! For many years, detectives have used fingerprints taken from crime scenes to help identify criminals. Now police can also use DNA "fingerprinting," or DNA profiling, to solve crimes. Except for identical siblings, no two people have the same DNA. Detectives can use samples of skin, hair, blood, and saliva to help identify criminals.

DNA is even helping free wrongly convicted people. Investigators are rechecking old evidence with new DNA technology. In some cases, they find that it was someone else's DNA evidence left at the crime scene.

Finding Family

DNA profiling is helping the military identify the badly damaged bodies of soldiers. Before the use of DNA, these soldiers were "unknown." Now DNA can be taken from a soldier's skeleton and matched to relatives on the mother's side of the family. This makes it possible for families to claim their lost soldiers.

THE HUMAN GENOME

• • •

In June of 2000, scientists announced that they had successfully mapped the human **genome**. What is a human genome? A genome is a complete set of genes. The human genome is the total set of genes found in humans.

The Human Genome Project began in 1990. With the help of computers and several laboratories across the country, the project finished ahead of schedule. We now have a "map" of where every human gene is located in the chromosomes.

Scientists are putting this information to good use. They are gaining a better understanding of how humans develop, how the human body works, and what can make things go wrong. Hopefully, knowledge about the human genome will lead to new treatments or prevention methods for many genetic disorders. Scientists have identified several genes that put people at high risk for problems like Alzheimer's disease and some cancers. If people know they have these genes, they can take steps to prevent the disease or at least get early treatment.

Closer Than You Think

Has anyone ever told you that you're as quiet as a mouse or you're behaving like an ape? They might have been right! About ninety percent of a human genome is the same as that of a mouse. Ninety-eight percent is identical to a chimpanzee's genome.

GENETIC COUNSELING

● ● ●

What if you carry a recessive gene for a disease and don't know it? What if you carry a dominant gene for an undesirable trait that you wouldn't wish to pass on to your children? Do you know what the chances are that your children will get certain characteristics present in your family? With today's advances in genetics, genetic counseling can help answer these questions and more. Genetic counselors are people who study family histories to provide information about future generations. They can help you understand your family's hereditary patterns and how they might affect you or your children in the future.

CLONING

● ● ●

Cloning is the creation of a new organism that is genetically identical to the organism from which it was produced. In other words, a clone has the exact same genes as its parent.

Have you ever taken a small part of a plant from a larger one, put it in water until it grew roots, and then planted it in soil where it grew into a new plant? That's cloning. When strawberry plants put out runners that grow into new strawberry plants, that's cloning. When a flatworm is cut into pieces and each piece regrows into a new worm, that's cloning. All asexual reproduction is, in fact, cloning.

Try This!

Clone a plant. You can start a new coleus or begonia plant by cutting a leaf from an old one. Plant the leaf in potting soil and water it. The new plant that grows will be a genetic clone of the original plant.

The cloning of organisms that normally reproduce sexually, however, is a new development in science. Because most animals and humans reproduce by combining the genes of both parents, the resulting offspring is not an exact genetic copy of anyone. They are not clones.

But scientists have recently developed ways to reproduce some of these animals so that a clone *is* produced. In 1996, scientists in Scotland shocked the world by successfully cloning a sheep named Dolly. Dolly made headlines around the world because she was the first mammal to be cloned from an adult cell. To create Dolly, researchers replaced the nucleus of an egg (reproductive) cell from one sheep with the nucleus from a cell from another sheep's udder. The egg was then implanted into a third sheep's uterus to develop. The clone then had the same DNA as the sheep from which the udder cell came.

First Sheep

Second Sheep

Nucleus replaced

New cell is activated with a jolt of electricity

Cell develops into an embryo

Third Sheep

Dolly

Dolly

Since Dolly, several other animals have been cloned. Mice, pigs, sheep, goats, monkeys, cats, and rabbits are among these animals. However, most of these clone offspring died or suffered from severe health problems. The question still remains as to whether humans can or should be successfully cloned in the future.

Many people see advantages to cloning. Researchers can use cloned animals to study genetic mutations and human diseases. Cloning could reproduce valuable animals with the same rare traits as the originals. Cloning may help save endangered species from becoming extinct. Italian scientists have already cloned an endangered animal—a wild sheep called a *mouflon*. Some scientists even dream of bringing back an already extinct species. Scientists have collected DNA from a woolly mammoth that was found frozen in the ice in Siberia. Some scientists believe that perhaps a cloned mammoth embryo might be able to grow in the womb of a female elephant. Animal cloning might also be used to save human lives. Cloned pigs might provide organs to transplant into humans. Cow clones might help manufacture life-saving medicines.

Dolly Obituary

In 2003, at the age of six, Dolly was put to sleep. She was suffering from lung cancer and crippling arthritis. Dolly was survived by six children (lambs) all born the "normal" way.

Failed Froggies

Frogs were cloned in the 1970s. Unfortunately, the tadpoles that developed would not grow into new frogs.

Despite the positive possibilities for cloning, some people have raised serious questions about the process. How many animals will be produced only to get sick and die? What effects might an animal organ have on the person who receives it? Should humans interfere with the natural cycle of life? Much more work with cloning must be done to address these and other concerns.

GENETIC ENGINEERING
● ● ●

Farmers have been improving crops for thousands of years. They save the seeds of the best plants year after year. They make sure the best plants pollinate the best plants. Through this "selective breeding," farmers can grow plants that are stronger, produce more, or taste better. Over many years, cauliflower, brussels sprouts, broccoli, and different colors of cabbage have all been developed from the original wild cabbage plant through selective breeding. But now genetics has opened up a whole new way to speed up the improvement of crops. It's called *genetic engineering.*

In genetic engineering, scientists combine pieces of DNA from different species. These "recombined" DNA molecules can be used to engineer, or produce, plants with desirable traits such as a resistance to disease or insects. By changing plant DNA in a laboratory, scientists can create plants that need less water, grow in poor soil, survive cold temperatures, and produce better-tasting fruit that lasts longer without spoiling. Crops that have been genetically altered are called GM, or genetically modified, crops.

The Genetic Engineering of Foods Debate

Advantages	Disadvantages
higher quantity	unknown health effects
better taste	could harm the environment
increased nutritional value	loss of variety within species
improved resistance to bugs and diseases	might be stressful or harmful to genetically engineered animals
less growing time	a few large food companies could take control of the food market
large populations could feed their people more easily	interferes with the natural life cycle

Genetic engineering can be done with animals and bacteria as well as plants. Chickens can be genetically improved to lay more eggs. Dogs and cats can be genetically reproduced with traits that make them more valuable for breeding. The genetic engineering of bacteria and other organisms can create medicines for human genetic disorders.

Cows Help with Clotting

Genetic engineering is used to manufacture medicine for people with hemophilia. Hemophilia is a genetic disorder that prevents blood from clotting. Scientists have inserted human genes into the cells of cows. The cows then produce the human blood-clotting protein in their milk. The milk can be used to treat people with hemophilia.

GENE THERAPY

● ● ●

Gene therapy is another current development in genetics. It involves inserting working copies of a gene directly into the cells of a person with a genetic disorder. The hope is that the "good" genes will take over for the "bad" genes and the disease or disorder will go away. Gene therapy is still in the experimental phase, but it is hoped that one day it will be a cure for many genetic disorders.

IN CONCLUSION

● ● ●

Genetic research and engineering are creating a scientific and medical revolution. They are dramatically changing the world and how we live. It's amazing how far genetics has come since Gregor Mendel's simple experiments with pea plants. Yet his basic understanding of heredity has led to a future of possibilities—one gene at a time.

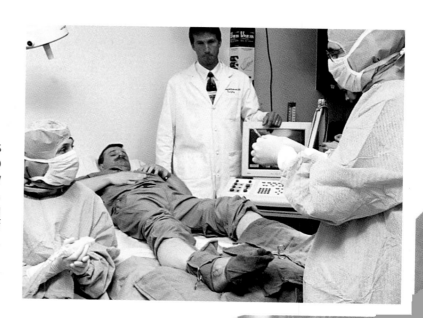

Donovan Decker was the first person to receive gene therapy for muscular dystrophy. Here he is preparing for an injection of healthy genes.

INTERNET CONNECTIONS AND RELATED READING FOR GENETICS
● ● ●

Internet sites

http://www.genecrc.org/site/ko/index_ko.htm
Go-Go Gene will introduce you to genes, chromosomes, and DNA at this "Kids Only" section of the Gene Cooperative Research Centre site. Refer to the glossary of genetics words, play some gene games, and use the links to explore heredity further.

http://www.dnaftb.org/dnaftb/1/concept/index.html
Review and extend your knowledge of genetics with the step-by-step information provided by the Dolan DNA Learning Center.

http://library.thinkquest.org/19037/heredity.html
Attend "The Gene School" to learn more about heredity traits, Punnett squares, Mendel, and modern uses of DNA.

http://kidshealth.org/kid/talk/qa/what_is_gene.html
It's all in the genes! Find out what that means with this Kids' Health explanation.

http://www.genetics.gsk.com/kids/index_kids.htm
Kids Genetics from GlaxoSmithKline has games to teach about genes, DNA, and heredity as well as fun factoids like banana split vaccines.

http://www.oneworld.org/penguin/genetics/home.html
Tiki the Penguin shares his knowledge and opinions on genetics and genetic engineering.

http://www.ornl.gov/sci/techresources/Human_Genome/home.shtml
This Human Genome Project site addresses many of the advancements made in the field of genetics, including gene therapy, cloning, genetic engineering, and much more.

Books

Francis Crick & James Watson: And the Building Blocks of Life by Edward Edelson. Biography of Francis Crick and James Watson, cowinners of a Nobel Prize for proposing the double helix structure of DNA. Oxford University Press, 1998. [RL 7 IL 6–12] (3995701 PB)

Plant Reproduction by Louise and Richard Spilsbury. Explores the reproductive processes of plants. Heinemann Library, 2003. [RL 3 IL 3–5] (3453601 PB)

Reproduction and Growth by Steve Parker. Offers students a comprehensive picture of reproduction and growth. Millbrook Press, 1998. [RL 5 IL 4–6] (3111206 HB)

•RL = Reading Level •IL = Interest Level
Perfection Learning's catalog numbers are included for your ordering convenience.
PB indicates paperback. HB indicates hardback.

Glossary

allele	(uh LEEL) one form of a gene
base	(bays) one of four molecules that make up DNA (see separate entries for *molecule* and *DNA*)
breeder	(BREED er) person who grows plants or raises animals under controlled conditions
cell	(sel) smallest unit of life that makes up all living things
chromosome	(KROH muh zohm) structure inside a cell's nucleus that carries genes (see separate entries for *cell* and *nucleus*)
cloning	(KLOHN ing) creation of a new organism with identical genes as the organism from which it was produced
codominant	(koh DAH muh nuhnt) type of gene that produces an observable trait mixed with another codominant trait
cytoplasm	(SEYET oh plaz uhm) gel-like substance found outside a cell's nucleus (see separate entries for *cell* and *nucleus*)
DNA	(dee en ay) molecule containing genetic information; deoxyribonucleic acid (see separate entry for *molecule*)
dominant	(DAH muh nuhnt) type of gene that controls which trait is always observable when present in an organism
embryo	(EM bree oh) animal or plant in the early stage of growth
environment	(en VEYE er muhnt) circumstances, objects, and conditions that surround someone
evolution	(ev uh LOO shuhn) development and change of all organisms over time

gene (jeen) unit of heredity that determines traits (see separate entry for *heredity*)

genetics (juh NET iks) science of heredity (see separate entry for *heredity*)

genome (JEE nohm) complete set of genes in an organism

heredity (huh RED uh tee) passing on of traits from one generation to another

molecule (MAHL uh kyoul) tiny particle of a substance made up of two or more units of matter

mutant (MYOU tent) offspring with a trait caused by gene mutation (see separate entry for *mutation*)

mutation (myou TAY shuhn) change in a gene or chromosome (see separate entry for *chromosome*)

nucleus (NOO klee uhs) part of a cell that controls the cell's activities (see separate entry for *cell*)

offspring (AWF spring) new organism; child

organism (OR guh niz uhm) living thing

protein (PROH teen) molecule in all cells that helps with growth, repair, and replacement of cells (see separate entries for *molecule* and *cell*)

recessive (ree SES iv) type of gene that is passed on but not observable when a dominant gene is present (see separate entry for *dominant*)

reproduction (ree pruh DUK shuhn) process by which a living thing produces more of itself

ribosome (REYE buh sohm) part of a cell that makes proteins (see separate entries for *cell* and *protein*)

trait (trayt) characteristic passed on through generations

Index